PHYSICAL CHANGES

Rebecca Kraft Rector

Enslow Publishing
101 W. 23rd Street
Suite 240
New York, NY 10011
USA

enslow.com

Words to Know

atom A tiny bit of matter.

chemical Having to do with chemistry.

chemistry The science that deals with properties of matter and how it forms and changes.

gas A kind of matter that has no permanent shape, like air.

liquid A kind of matter that can move freely, like water.

physical Having to do with being able to be touched or seen.

properties The qualities or features of something.

solid A kind of matter that is firm and keeps its shape.

CONTENTS

WORDS TO KNOW 2
WHAT IS MATTER? 5
COMMON FORMS OF MATTER 7
PROPERTIES OF MATTER 9
CHEMICAL CHANGES 11
PHYSICAL CHANGES 13
FAST AND SLOW PHYSICAL CHANGES 15
PHYSICAL CHANGE BY HEAT AND COLD 17
PHYSICAL CHANGE BY MIXING 19
PHYSICAL CHANGES ARE EVERYWHERE 21
ACTIVITY: WHAT'S THE MATTER? 22
LEARN MORE 24
INDEX 24

What Is Matter?

Matter is everything around you. All things are made of matter. Tiny bits of matter are called atoms. Atoms join together to make molecules.

Fast Fact

Even people are made of matter.

We can see solids and liquids like a kite and the ocean. Often we cannot see gas, like the wind that blows the kite.

Common Forms of Matter

Matter has different forms. Matter can be solid. A book is a solid. Matter can be liquid. Milk is a liquid. Matter can be a gas. Oxygen is a gas.

Fast Fact

Atoms are packed tightly together in a solid.

We can see and feel physical properties. The kitten is soft, and the yarn is orange.

Properties of Matter

Properties tell about matter. Physical properties tell how it acts, looks, and feels. Examples include size and softness. Chemical properties let matter change. An example is being able to rust.

**Eating a cookie causes a chemical change.
Your body changes the cookie into different materials.**

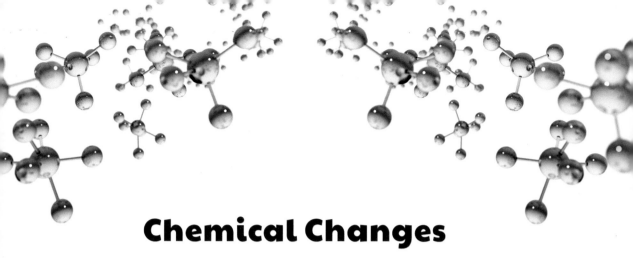

Chemical Changes

Matter can make a chemical change. Dough bakes. It changes into bread. The atoms are joined in a different way. A new material is formed.

Fast Fact

Usually you cannot undo a chemical change.

Chopping wood causes a physical change. The wood looks different, but it is still the same material.

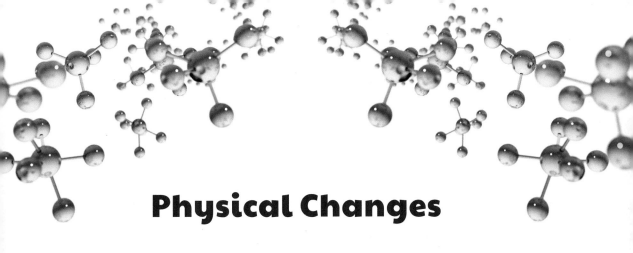

Physical Changes

Matter can make a physical change. A plate breaks. It looks different. The pieces are smaller. They are different shapes. But they are still made of the same material.

Fast Fact
Physical changes can often be changed back.

Trimming a dog's nails is a fast physical change.

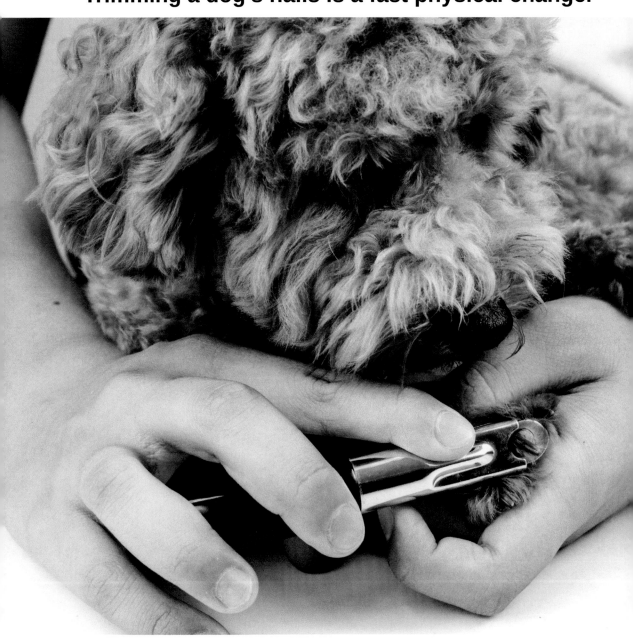

Fast and Slow Physical Changes

Physical changes can be fast or slow. Shredding paper is fast. The paper is smaller. It is still paper. Water slowly wears down rock. It is still rock.

Water can change forms. When water vapor hits cool glass, it changes to a liquid.

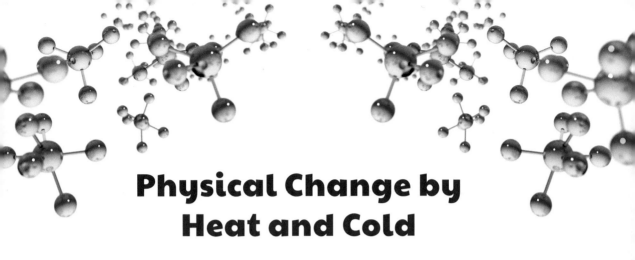

Physical Change by Heat and Cold

Cold and heat can change the form of matter. Water is a liquid. Freezing water makes solid ice. Boiling water makes gassy steam. These are all forms of water.

Fast Fact

Melting candles have a physical change in shape.

Mixing different kinds of solid foods causes a physical change.

Physical Change by Mixing

Make a pile of red buttons. Mix in white buttons. The pile looks different. But a new material is not made. It is a physical change.

FAST FACT

Mixing paint colors together makes a physical change.

Making a snowman only changes the snow's form. It is a physical change.

Physical Changes Are Everywhere

Nature makes changes. The sun melts ice cream. The ocean moves sand. People make changes. People cut grass. They paint a room. Physical changes are all around you.

Activity
What's the Matter?

Watch matter change! Let's get started!

MATERIALS

Water
Ice cube tray
Freezer
Zip-top plastic baggie
Masking tape
Sunny window

Procedure:

Step 1: Fill the ice cube tray with water.

Step 2: Put the tray in the freezer. Wait for ice cubes to form.

Step 3: Put three ice cubes in the baggie. Seal tightly.

Step 4: Tape the baggie inside a sunny window.

Step 5: Watch what happens after several hours.

Liquid water changed to solid ice and back again! Are there drops of water in the baggie? Water vapor (gas) is changing back to liquid.

It's easy to make physical changes in water.

Learn More

Books

Haelle, Tara. *Matter Changing States*. Vero Beach, FL: Rourke, 2018.

Rompella, Natalie. *Experiments in Material and Matter with Toys and Everyday Stuff*. North Mankato, MN: Capstone, 2016.

Sjonger, Rebecca. *Changing Matter in My Makerspace*. New York, NY: Crabtree, 2018.

Websites

Scholastic
studyjams.scholastic.com / studyjams / jams / science / matter / changes-of-matter.htm
Watch this fun video and discover more about changes in matter.

Science Kids
www.sciencekids.co.nz / gamesactivities / reversiblechanges.html
Learn more about changes in matter with these fun games.

Index

atoms, 5, 7, 11
chemical changes, 11
chemical properties, 9
cold, 17
fast changes, 15
gas, 7, 17
heat, 17
liquid, 7, 17
matter, 5, 7, 9, 11, 13, 17
mixing, 19
molecules, 5
physical changes, 13, 15, 17, 19, 21
physical properties, 9
slow changes, 15
solid, 7, 17

Published in 2020 by Enslow Publishing, LLC.
101 W. 23rd Street, Suite 340, New York, NY 10011

Copyright © 2020 by Enslow Publishing, LLC.

All rights reserved.

No part of this book may be reproduced by any means without the written permission of the publisher.

Library of Congress Cataloging-in-Publication Data

Names: Rector, Rebecca Kraft, author.
Title: Physical changes / Rebecca Kraft Rector.
Description: New York : Enslow Publishing, 2020. | Series: Let's learn about matter | Audience: K to grade 4. | Includes bibliographical references and index.
Identifiers: LCCN 2018045962| ISBN 9781978507586 (library bound) | ISBN 9781978509139 (pbk.) | ISBN 9781978509146 (6 pack)
Subjects: LCSH: Matter—Properties—Juvenile literature. | Chemistry, Physical and theoretical—Juvenile literature.
Classification: LCC QC173.36 .R43 2020 | DDC 541—dc23

LC record available at https://lccn.loc.gov/2018045962

Printed in the United States of America

To Our Readers: We have done our best to make sure all website addresses in this book were active and appropriate when we went to press. However, the author and the publisher have no control over and assume no liability for the material available on those websites or on any websites they may link to. Any comments or suggestions can be sent by e-mail to customerservice@enslow.com.

Photo Credits: Cover, p. 1 Captured by Nicole/Shutterstock.com; p. 4 Yuganov Konstantin/Shutterstock.com; p. 6 NadyaEugene/Shutterstock.com; p. 8 Vagengeim/Shutterstock.com; p. 10 Thomas M Perkins/Shutterstock.com; p. 12 Jan Faukner/Shutterstock.com; p. 14 ThamKC/Shutterstock.com; p. 16 yenphoto24/Shutterstock.com; p. 18 Dorling Kindersley ltd/Alamy Stock Photo; p. 20 Suzanne Tucker/Shutterstock.com; p. 23 al1962/Shutterstock.com; interior design elements (ice cube) BlueRingMedia/Shutterstock.com, (molecules) 123dartist/Shutterstock.com.